The Practically Cheating Statistics Handbook

TI-83 Companion Guide

S. Deviant, MAT

StatisticsHowTo.com

info@statisticshowto.com

Kenrose Media

Jacksonville, FL

Editor in Chief: *Pete Michaud*

Kenrose Media © 2010 Kenrose Media, LLC

Printed in the United States of America

ISBN 1-45-379816-1

Contents

Basics

Mean and Median

Finding the mean or median from a list of data can be accomplished in two ways: by entering a list of data, or by using the home screen to type the commands. Using the list feature is just as easy as entering the data onto the home screen, and it has the added advantage that you can use the data for other purposes after you have calculated your mean and median (for example, you might want to create a histogram).

Sample problem: Find the mean and the median for the height of the top 20 buildings in NYC. The heights, (in feet) are: 1250, 1200, 1046, 1046, 952, 927, 915, 861, 850, 814, 813, 809, 808, 806, 792, 778, 757, 755, 752, and 750.

Step 1: Enter the above data into a list. Press the `STAT` button and then press `ENTER`. Enter the first number (1250), and then press `ENTER`. Continue entering numbers, pressing the `ENTER` button after each entry.

Step 2: Press the `STAT` button.

Step 3: Press 🠖 to highlight **Calc**.

Step 4: Press `ENTER` to choose **1-Var Stats**.

Step 5: Press `ENTER` again. The calculator will return the mean, x. For this list of data, the mean is **877.667** feet (rounded to 3 decimal places). The standard deviation is Sx=**147.022**.

Step 6: Arrow down until you see **Med**. This is the median; for the above data, the median is **813** feet.

Interquartile Range

The interquartile range is the distance between the top of the lower quartile and the bottom of the top quartile.

Sample problem: Find the interquartile range for the heights of the top 10 buildings in the world. The heights, (in feet) are: 2717, 2063, 2001, 1815, 1516, 1503, 1482, 1377, 1312, 1272.

Step 1: Enter the above data into a list. Press the STAT button and then press ENTER. Enter the first number (2717), and then press ENTER. Continue entering numbers, pressing ENTER after each entry.

Step 2: Press STAT.

Step 3: Press → (the arrow keys are located at the top right of the keypad) to select **Calc**.

Step 4: Press ENTER to highlight **1-Var Stats**.

Step 5: Press ENTER again to bring up a list of stats.

Step 6: Scroll down the list with the arrow keys to find Q1 and Q3. Subtract Q3 from Q1 to find the **interquartile range** (which in this set of numbers would be **624** feet).

Variance

The variance is simply the standard deviation squared. You could find the standard deviation for a list of data and square the result, but you won't get an accurate answer unless you square the entire answer, including *all* of the significant digits (not, for example, just the two I rounded to in the standard deviation question). Here's how to get the TI-83 variance without you having to perform a manual calculation:

Sample problem: Find the variance for the heights of the top 12 buildings in London, England. The heights, (in feet) are: 800, 720, 655, 655, 625, 600, 590, 529, 513, 502, 502, 502.

Step 1: Enter the above data into a list. Press the STAT button and then press ENTER. Enter the first number (800), and then press ENTER. Continue entering numbers, pressing ENTER after each entry.

Step 2: Press STAT. Press → to select **Calc**.

Step 3: Press ENTER to highlight **1-Var Stats**. Press ENTER again to bring up a list of stats.

Step 4: Press VARS 5 to bring up a list of the available Statistics variables.

Step 5: Press 3 to select "Sx" which is our standard deviation.

Step 6: Press x^2, then ENTER to display the variance, which is **9326.628788**.

Standard Deviation

Standard deviations give us an idea of how much data is contained within a certain part of a distribution curve. This information is especially useful when using normal distribution curves (which you'll, no doubt, learn about during your statistics course).

Sample problem: Find the standard deviation for the heights of the top 12 buildings in London, England. The heights, (in feet) are: 800, 720, 655, 655, 625, 600, 590, 529, 513, 502, 502, 502.

Step 1: Enter the above data into a list. Press the STAT button and then press ENTER. Enter the first number (800), and then press ENTER. Continue entering numbers, pressing ENTER after each entry.

Step 2: Press STAT.

Step 3: Press → (the arrow keys are located at the top right of the keypad) to select **Calc**.

Step 4: Press ENTER to highlight **1-Var Stats**.

Step 5: Press ENTER again to bring up a list of stats. The standard deviation (Sx) for the above list of data is **96.57** feet (rounded to 2 decimal places).

Graphs

Box Plot

Sample problem: You have a list of IQ scores for a gifted classroom in a particular elementary school. The IQ scores are: 118, 123, 124, 125, 127, 128, 129, 130, 130, 133, 136, 138, 141, 142, 149, 150, 154. That list doesn't tell you much about anything. Create a box plot to make sense of this data.

Step 1: Press STAT ENTER, to edit L1.

Step 2: Enter the data from the problem into the list:

1	1	8	ENTER
1	2	3	ENTER
1	2	4	ENTER
1	2	5	ENTER
1	2	7	ENTER
1	2	8	ENTER
1	2	9	ENTER
1	3	0	ENTER
1	3	0	ENTER
1	3	3	ENTER
1	3	6	ENTER
1	3	8	ENTER
1	4	1	ENTER
1	4	2	ENTER
1	4	9	ENTER
1	5	0	ENTER
1	5	4	ENTER

Step 3: Press [2ND] [Y=], to access the **Stat Plot** menu.

Step 4: Press [ENTER] [ENTER] to turn on **Plot1**.

Step 5: Arrow down to **Type**, which has 6 icons to the right of it. Highlight the bottom middle icon, which looks like a syringe with two plungers, and press [ENTER] to select it.

Step 6: Make sure the **XList** entry reads "L_1". If it doesn't, arrow down to it, Press [CLEAR] [2ND] [1].

Step 7: Press [GRAPH]. You should see your Box plot!

Tip: If when you press [GRAPH], you see the message "Err: Stat", or you just don't see a box plot like you expect to, then press [WINDOW], and try different settings. Especially try changing the **Xscl** (X Scale) item to a larger value.

Cumulative Frequency Table

A cumulative frequency table shows shows the sum of data up to a specified point. A TI 83 can draw a cumulative frequency table, using values already placed in a list. Just type in the numbers, press a button, and the calculator does the rest.

Sample problem: Construct a cumulative frequency table for the grades of school children in Rock Hill elementary school (grades k-5). The grades are A (90+), B (80+), C (70+), and D (60+). The number of students achieving those grades are: A (20), B (21) C (52) and D (29).

Step 1: Enter the data list L1. Press the 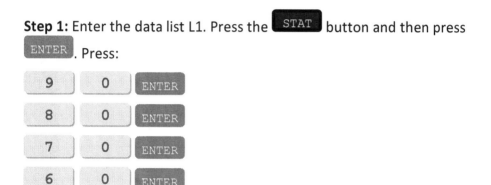 STAT button and then press ENTER. Press:

9	0	ENTER
8	0	ENTER
7	0	ENTER
6	0	ENTER

Step 2: Scroll to the right to list L2. Enter the next column's values:

2	0	ENTER
2	1	ENTER
5	2	ENTER
2	9	ENTER

Step 3: Scroll to the right and up to highlight L3.

Step 4: Press `2ND` `STAT` to reach the **LIST** menu.

Step 5: Scroll to the right to highlight **OPS** then press `6` for **cumSum**.

Step 6: Press `2ND` `1` to get L1, then press `)`.

Step 7: Press `ENTER`. The TI-83 will sum the cumulative data in the L2 column and replace the results in L3, giving you a cumulative distribution table.

Tip: To clear the data from a list, place the cursor on the list name (for example, L1), press `CLEAR`, and then `ENTER`.

Frequency Chart or Histogram

A histogram is one of the most popular ways the display a frequency distribution. A series of vertical rectangles fit together on a chart. The histogram is an important tool for exploratory data analysis, and can be plotted on a TI 83 graphing calculator in less time than it would take you to get out a sheet of paper, a pencil, and a ruler.

Sample problem: draw a histogram for the following recent test scores in a statistics class: 45, 67, 68, 69, 74, 76, 75, 77, 79, 84, 86, 90.

Step 1: Press STAT ENTER , to edit L1.

Step 2: Enter the data from the problem into the list:

4	5	ENTER
6	7	ENTER
6	8	ENTER
6	9	ENTER
7	4	ENTER
7	6	ENTER
7	5	ENTER
7	7	ENTER
7	9	ENTER
8	4	ENTER
8	6	ENTER
9	0	ENTER

Step 3: Press `2ND` `Y=`, to access the **Stat Plot** menu.

Step 4: Press `ENTER` `ENTER` to turn on **Plot1**.

Step 5: Arrow down to **Type**, which has 6 icons to the right of it. Highlight the top right icon, which looks like a histogram, and press `ENTER` to select it.

Step 6: Make sure the **XList** entry reads "L_1". If it doesn't, arrow down to it, `CLEAR` `2ND` `1`.

Step 7: Press `GRAPH`. You should see your Histogram!

Tip: If when you press `GRAPH`, you see the message "Err: Stat", or you just don't see a histogram like you expect to, then Press `WINDOW`, and try different settings. Especially try changing the **Xscl** (X Scale) item to a larger value.

Scatter Plot

In order to graph a TI-83 scatter plot, you'll need a set of "bivariate" data. Bivariate data means you have both X and Y values, for example, weight (your X) and height (your Y). The XY values can be in two separate lists, or they can be written as XY cordinates (x,y).

Sample problem: Create a TI-83 scatter plot for the following coordinates (2,3), (4,4), (6,9), (8,11), and (10,12).

Step 1: Press STAT ENTER to enter the lists screen. If you already have data in L1 or L2, clear the data: move to cursor onto L1, press CLEAR and then ENTER. Repeat for L2.

Step 2: *Enter your x-variables, one at a time.* Follow each number by pressing the ENTER key. For our list, you would enter:

Step 3: Use the ➡ key to scroll across to the next column, L2.

Step 4: Enter your y-variables, one at a time. Follow each number by pressing the enter key. For our list, you would enter:

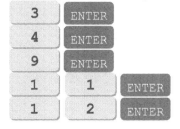

Step 5: Press `2ND` and then press `Y=` (Statplot).

Step 6: Press `ENTER` to enter StatPlots for Plot1.

Step 7: Press `ENTER` to turn Plot1 "ON."

Step 8: Arrow down to the next line ("Type") and highlight the scatter plot (the first image).

Step 9: Arrow down to "Xlist." If "L1" isn't showing, press `2ND` and `1`. Arrow down to "Ylist." If "L2" isn't showing, press `2ND` and `2`.

Step 10: Press `ZOOM` and `9`. This should bring up a scatter plot on your screen.

Tip: Hit `TRACE` and press the right and left arrow buttons to move from point to point, displaying the XY values for those points.

Binomial Probability Distribution

Combinations

Sample problem: If there are 5 people, Barb, Sue, Jan, Jim, and Rob, and only three will be chosen for the new Parent Teacher Association, how many combinations are possible for the committee?

Step 1: Enter the number of items on the home screen (the main input area) that you have to choose from. In the above example, you have 5 people, so press ⬚ 5 ⬚.

Step 2: Press the ⬚ MATH ⬚ button.

Step 3: Press ➡ to tab across to **PRB**.

Step 4: Press ⬚ 3 ⬚. This inserts **nCr** onto the home screen.

Step 5: Enter the number of items you want to choose. In the above example, you want to choose 3, so press ⬚ 3 ⬚.

Step 6: Press ⬚ ENTER ⬚. The calculator will return the result: **10**.

Permutations

Sample problem: If there are 5 people, Barb, Sue, Jan, Jim, and Rob, and three will be chosen for the new Parent Teacher Association. The first person picked will be the president, then the vice president, then the secretary. How many permutations are possible for the committee?

Step 1: Enter the number of items on the home screen (the main input area) that you have to choose from. In the above example, you have 5 people, so press 5 .

Step 2: Press the MATH button.

Step 3: Press → to tab across to **PRB**.

Step 4: Press 2 . This inserts **nPr** onto the home screen.

Step 5: Enter the number of items you want to choose. In the above example, you want to choose 3, so press 3 .

Step 6: Press ENTER . The calculator will return the result: **60**.

Mean and Standard Deviation for a Binomial Probability Distribution

Binomial probability distributions are used when dealing with two outcomes from a fixed number of independent trials. As a simple example, you might want to figure out the probability of a voter voting yes or no on a certain ballot proposition. P(x) can quickly be calculated on the TI 83 graphing calculator, which has all of the binomial probability tables built into it.

Sample problem: Find the mean and standard deviation for a binomial distribution with n = 5 and p = 0.12.

Mean
Step 1: Multiply n by p:

5	×	.	1	2	ENTER

= .6

Hey, that was easy!

Standard Deviation
Step 1: Subtract p from 1 to find q.

1	−	.	1	2	ENTER

= .88

Step 2: Multiply n times p times q.

5	×	.	1	2	×	.	8

8	ENTER

= .528

Step 3: Find the square root of the answer from Step 2.

2ND	x^2	.	5	2	8	ENTER

.727 (rounded to 3 decimal places).

Bionomial Probability: BinomCDF

The TI-83 BinomPDF and TI-83 BinomCDF functions can assist you in solving binomial probability questions in seconds. The difference between the two functions is that one (BinomPDF) is for a single number (for example, three tosses of a coin), while the other (BinomCDF) is a cumulative probability (for example, 0 to 3 tosses of a coin).

Sample problem: Bob's batting average is .321. If he steps up to the plate four times, find the probability that he gets up to three hits.

Step 1: Press `2ND` `VARS` to get the **DISTR** option. Scroll down to **A:BinomCDF** and then press `ENTER`.

Step 2: Enter the number of trials. Bob bats four times, so the number of trials is **4**. Press `4` followed by `,`.

Step 3: Enter the Probability of Success, **P**. Bob has a batting average of **.321**, so enter `.` `3` `2` `1`, then press `,`.

Step 4: Enter the X value. We want to know what the probability is that Bob will get up to three hits, so enter `3` and then press `)`.

Step 5: Press `ENTER` for the result. Bob's chances of getting up to three hits is **.989**.

Tip: Instead of scrolling to **A:BinomCDF** you can press `ALPHA` `MATH`.

Bionomial Probability: BinomPDF

Sample problem: Sue has a batting average of .270. If she is at bat three times, what is the probability that she gets exactly two hits?

Step 1: Press **2ND** **VARS** to get the **DISTR** option. Press **0** for **binomPDF**.

Step 2: Enter the number of trials from the question. Sue bats three times, so the number of trials is **3**. Press **3** and hit the [,] key.

Step 3: Enter the Probability of Success, **P**. Sue's batting average is **.240**, so enter **.** **2** **7** **0**, then hit the [,] key.

Step 4: Enter the X value. We want to know Sue's chances of getting exactly two hits, so enter **2** and then press [ )].

Step 5: Press **ENTER** for the result. Sue's chances of getting exactly two hits is **.159651**.

Warning: **BinomialPdf** is an exact probability for one value of x. If you want to find a cumulative probability (for example, what are John's chances of getting 0 or 1 hits?) you will need the use the BinomialCdf function.

Normal Distributions

Central Limit Theorem

The TI 83 calculator has a built in function that can help you calculator probabilities of central theorem word problems. The function, normalcdf, requires you to enter a lower bound, upper bound, mean, and standard deviation.

Sample problem: A fertilizer company manufactures organic fertilizer in 10 pound bags with a standard deviation of 1.25 pounds per bag. What is the probability that a random sample of 150 bags will have a mean between 9 and 9.5 pounds?

Step 1: Press  .

Step 2: Enter your variables (lower bound, upper bound, mean, and standard deviation):

Step 3: Press ENTER . This returns the probability of **.159**, or **15.9%**.

Tip: Sampling distributions require that the standard deviation of the mean is σ / √(n), so make sure you enter that as the standard deviation.

Normally Distributed Probability

Sample problem: A group of students taking end of semester statistics exams at a certain college have a mean score of 75 and a standard deviation of 5 points. What is the probability that a given student will score between 90 and 100 points?

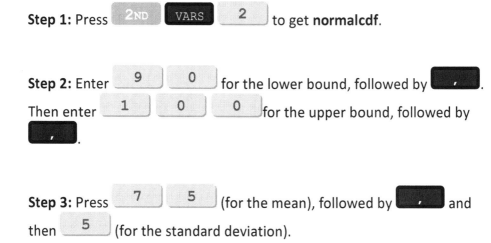

Step 1: Press 2ND VARS 2 to get **normalcdf**.

Step 2: Enter 9 0 for the lower bound, followed by ,. Then enter 1 0 0 for the upper bound, followed by ,.

Step 3: Press 7 5 (for the mean), followed by , and then 5 (for the standard deviation).

Step 4: Close the argument list with a). Your display should now read **normalcdf(90, 100, 75, 5)**. Now press ENTER. The calculator returns the probability, which in this case is **0.00135**, or **1.35%** (rounded to two decimal places).

Normal Probability Density Function

The TI-83 **normalPDF** function, accessible from the **DISTR** menu will calculate the normal probability density function, given the mean (μ) and standard deviation (σ).

Sample Problem: Graph a bell curve with a mean of 100 and standard deviation of 15.

Step 1: Press `Y=`.

Step 2: Press `2ND` `VARS` `1` to get **normalPDF**.

Step 3: Press `X,T,Θ,n` `,` `1` `0` `0` `,` `1` `5` `)`.

Step 4: Press `WINDOW`.

Step 5: Change the window values to the following:
Xmin=0
Xmax=150
Xscl=1
Ymin=0
Ymax=.05

Step 6: Press `GRAPH` to see your graph.

Z Scores

Find a Critical Value

A critical value in hypothesis testing separates the region where the hypothesis will be rejected from the region where the hypothesis will not be rejected. With a little arithmetic, you can look up a critical value in a z-table, or you can use the **InvNorm** function on the TI-83 graphing calculator.

Sample problem: An end of semester exam is normally distributed with a mean of 85 and a standard deviation of 10. Find the z score at the 90th percentile.

Step 1: Press . This displays **InvNorm(** on the home screen.

Step 2: Type . Your display should read **InvNorm(.9, 85, 1, 0)**.

Step 3: Press ENTER. This returns **97.82**. That means that 90% of students will have scores below 97.82.

Tip: The first entry on **InvNorm** should be a number between 0 and 1.

Hypothesis Testing

Hypothesis Test on a Mean

Researchers use data from a population sample to figure out when a hypothesis is true or not. If the sample data does not agree with the hypothesis, then the hypothesis is rejected. The null hypothesis (H_0) states that the observations occur by chance. The alternate hypothesis (H_1) is just the opposite, and states that the observations were influenced by a non-random cause.

Sample problem: A sample of 200 people has a mean age of 21 with a population standard deviation (σ) of 5. Test the hypothesis that the population mean is 18.9 at $\alpha = 0.05$.

Step 1: State the null hypothesis. In this case, the null hypothesis is that the population mean is 18.9, so we write:
$H_0: \mu = 18.9$

Step 2: State the alternative hypothesis. We want to know if our sample, which has a mean of 21 instead of 18.9, really is different from the population, therefore our alternate hypothesis:
$H_1: \mu \neq 18.9$

Step 3: Press STAT then press → → to select **TESTS**.

Step 4: Press 1 to select **1:Z-Test…**. Press ENTER.

Step 5: Press → to select **Stats**.

Step 6: Enter the data from the problem:
μ_0: 18.9
σ: 5
x: 21
n: 200
μ: $\neq \mu_0$

Step 7: Arrow down to **Calculate** and press ENTER . The calculator shows the p-value:

$p = 2.87 \times 10^{-9}$

This is smaller than our alpha value of .05. That means we should **reject the null hypothesis**, the average age of the population is *not* 18.9 years old.

Confidence Intervals

Confidence Intervals for a Proportion

Confidence intervals are one way you can decide how accurate data is. The higher the confidence, the better the estimate (a small range of possible values): a 98% confidence interval is going to give you a better estimate of the unknown parameter than a 90% confidence interval.

Sample problem: For x = 20, and n = 24, construct a 98% confidence interval for p, the true population proportion.

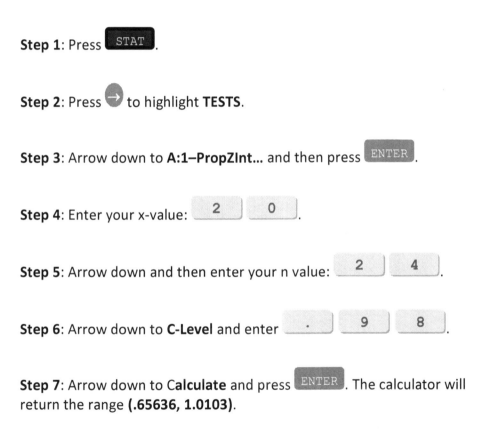

Step 1: Press STAT.

Step 2: Press → to highlight **TESTS**.

Step 3: Arrow down to **A:1–PropZInt...** and then press ENTER.

Step 4: Enter your x-value: 2 0.

Step 5: Arrow down and then enter your n value: 2 4.

Step 6: Arrow down to **C-Level** and enter . 9 8.

Step 7: Arrow down to **Calculate** and press ENTER. The calculator will return the range **(.65636, 1.0103)**.

Tip: Instead of arrowing down to select **A:1–PropZInt...**, press ALPHA and MATH instead.

Confidence Interval for a Mean (Known Standard Deviation)

The only difference between having statistics such as mean, or standard deviation, versus the raw data, is that you will have to enter the data into a list in order to perform the calculation.

Sample problem: 40 items are sampled from a normally distributed population with a sample mean (x) of 22.1 and a population standard deviation (σ) of 12.8. Construct a 98% confidence interval for the true population mean.

Step 1: Press `STAT`, then → to **TESTS**. Press `ENTER`.

Step 2: Press `7` for **Z Interval**.

Step 3: Arrow over to **Stats** on the **Inpt** line and press `ENTER` to highlight and move to the next line, σ.

Step 4: Enter `1` `2` `.` `8`, then arrow down to x.

Step 5: Enter `2` `2` `.` `1`, then arrow down to "n."

Step 6: Enter `4` `0`, then arrow down to "C-Level."

Step 7: Enter `.` `9` `8`. Arrow down to **Calculate** and then press `ENTER`. The calculator will give you the result of (17.392, 26.808) meaning that your 98% confidence interval is **17.392 to 26.808**.

T Scores

How to Find a T Distribution

The T distribution (sometimes called Student's T distribution) is actually a set of distributions, differentiated by the degrees of freedom (df). Take a look at a traditional textbook T table, and you'll actually find *many* T tables, which can be a little overwhelming. Instead of poring over tables, you can use a TI 83 graphing calculator to assist you in finding T distribution values. You might be asked to find the area under a T curve, or (like Z scores), you might be given a certain area and asked to find the T score.

Sample problem: Find the area under a T curve with degrees of freedom 10 for P(1 ≤ X ≤ 2).

Step 1: Press **2ND** **VARS** **5** to select **tcdf(**.

Step 2: Enter the lower, and upper bounds, and the degrees of freedom:

Your screen should now read **tcdf(1,2,10)**

Step 3: Press **ENTER**. The answer is **.133752549**, or about **13.38%**.

Linear Regression

Linear Regression

You can perform linear regression with a TI-83 in the time it takes you to input a few variables into a list. Linear regression will only give you a reasonable result if your data looks like a line on a scatter plot, so before you find the equation for a linear regression line you may want to view the data on a scatter plot first. See the Graphs chapter in this book to find out how to create a scatter plot.

Sample problem: Find a linear regression equation (of the form $y = ax + b$) for x-values of 1, 2, 3, 4, 5 and y-values of 3, 9, 27, 64, and 102.

Step 1: Press STAT ENTER to enter the lists screen. If you already have data in L1 or L2, clear the data: move to cursor onto L1, press CLEAR and then ENTER. Repeat for L2.

Step 2: *Enter your x-variables, one at a time.* Press:

Step 3: Use the ➔ key to scroll across to the next column, L2.

Step 4: Enter your y-variables, one at a time. Press:

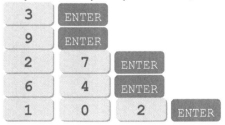

Step 5: Press the STAT button, then use the scroll key to highlight **CALC**. Press ENTER.

Step 6: Press 4 to choose **LinReg(ax+b)**. Press ENTER. The TI-83 will return the variables needed for the equation. Just insert the given variables (a, b) into the equation for linear regression (y=ax+b). For the above data, this is **y = 25.3x − 34.9**.

Two Populations

Confidence Intervals for Two Populations

Statistics about two populations is incredibly important for a variety of research areas. For example, if there's a new drug being tested for diabetes, researchers might be interested in comparing the mean blood glucose level of the new drug takers versus the mean blood glucose level of a control group. The confidence interval (CI) for the difference between the two population means is used to assist researchers in questions such as these. The TI 83 allows you to find a CI for the difference between two means in a matter of a few keystrokes.

Sample problem: Find a 98% confidence interval for the difference in means for two normally distributed populations with the following characteristics:

$x_1 = 88.5$
$\sigma_1 = 15.8$
$n_1 = 38$

$x_2 = 74.5$
$\sigma_2 = 12.3$
$n_1 = 48$

Step 1: Press STAT , then → highlight **TESTS**.

Step 2: Press 9 to select **2-SampZInt....**

Step 3: Enter the values from the problem into the appropriate rows, using the down arrow to switch between rows as you complete them.

Step 4: Press ↓ to select **Calculate** then press ENTER. The answer displayed is **(6.7467, 21.253)**. We're 98% sure that the difference between the two means is between 6.7467 and 21.253.

20444325R00027

Made in the USA
Lexington, KY
04 February 2013